SOIL SCIENCE
FOR BEGINNERS

BEST SOLUTION

Transforming Wastelands into Bountiful Fields: Embracing Sustainable Practices for Healthy Soil and Abundant Harvests is simple

TABLE CONTENT

INTRODUCTION	**1**
WHAT IS SOIL?	**3**
Physical characteristics of soil	4
Soil texture, structure, drainage characteristics	5
Soil colour	8
Determine soil drainage	9
Inorganic component of soils	10
Soil texture	11
Soil textural class	15
Organic component of soil	21
Soil organisms	24
SOIL ACIDITY	**27**

	WHAT CAUSES SOIL ACIDITY?	28
	HOW ACIDITY AFFECTS PLANT GROWTH	29
	Soil pH as a measure of acidity	31
COMPOSTING FOR SOIL		**38**
	Composting benefits	40
	What to compost	41
	How to compost step-by-step	45
A STEP-BY-STEP GUIDE TO GROWING MICROGREENS AND HERBS INDOORS		**55**
WHAT IS THE BEST SOIL TO USE FOR PLANTING FLOWERS?		**65**
	Loam Soil	67
	Silt Soil	69
	Clay Soil	73

INTRODUCTION

Delve into the fascinating world of soil science and discover the hidden wonders beneath our feet. This comprehensive beginner's guide will transform you into a soil aficionado, equipping you with the knowledge to understand, appreciate, and nurture the very foundation of life on Earth.

Embark on a journey into the intricate world of soil formation, exploring the physical, chemical, and biological processes that shape this essential resource. Unravel the mysteries of soil texture, structure, and composition, and gain insights into the diverse roles that soil plays in supporting plant growth, regulating water flow, and maintaining a healthy ecosystem.

As you delve deeper, you'll uncover the fascinating world of soil organisms, from microscopic bacteria to earthworms, and learn how their interactions contribute to soil fertility and health. Discover the importance of organic matter and its role in enhancing soil quality, and explore sustainable practices for maintaining soil health in the face of modern challenges.

INTRODUCTION

Whether you're a budding gardener, an enthusiastic farmer, or simply curious about the world around you, this guide will provide you with the essential knowledge to appreciate the wonders of soil science. With a deeper understanding of soil, you'll be empowered to make informed decisions that promote soil health and sustainability, ensuring a thriving future for our planet and its inhabitants.

Embrace the wonders of soil science and unlock the secrets of the Earth's most precious resource!

What is soil?

Soil is the loose surface material that covers most land. It consists of inorganic particles and organic matter. Soil provides the structural support for plants used in agriculture and is also their source of water and nutrients.

Soils vary greatly in their chemical and physical properties. Processes such as leaching, weathering and microbial activity combine to make a whole range of different soil types. Each type has particular strengths and weaknesses for agricultural production.

Physical characteristics of soil

The physical characteristics of soil include all the aspects that you can see and touch, such as:
- texture
- colour
- depth
- structure
- porosity (the space between the particles)
- stone content.

Good soil structure contributes to soil and plant health, allowing water and air movement into and through the soil profile. Soil stores water for plant growth and supports machine and animal traffic.

While some soils are naturally better structured than others, some physical characteristics of soils can be changed with good management.

It is important to monitor the physical characteristics of soil to understand soil conditions.

It is also important to ensure that management practices are not contributing to the decline of the soil.

An example of this is excessive traffic causing compaction and reducing the amount of macropores, or spaces between the aggregates, therefore reducing the amount of air and water into and through the soil.

Soil texture, structure, drainage characteristics

Soil aggregate

The combination of mineral fractions (gravel, sand, silt and clay particles) and organic matter fraction give soil its texture. Texture grades depend upon the amount of clay, sand, silt and organic matter present.

Sandy soil

The solid part of the soil is made up of particles such as organic matter, silt, sand and clay which form aggregates. Aggregates are held together by clay particles and organic matter. Organic matter is one of the major cementing agents for soil aggregates. The size and shape of aggregates give soil a characteristic called soil structure. Soil structure influences plant growth by affecting the movement of water, air and nutrients to plants. Sandy soils have little or no structure but are often free draining.

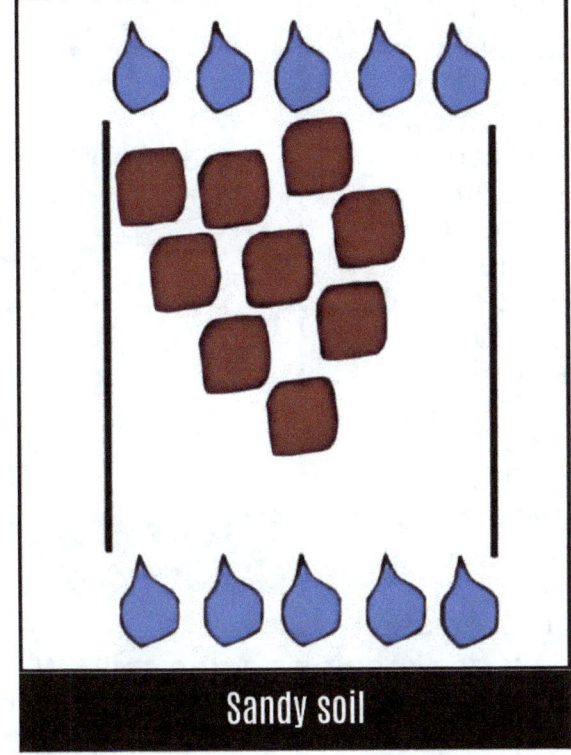

Sandy soil

With higher clay content, the soil structural strength increases, but its drainage ability often decreases.

Heavy clays can hold large amounts of water and, as infiltration rates are slow, they tend not to be well drained, unlike sand or loam soils with no or a lower clay content.
The number of soil pores and the pore size relate to the drainage capacity of the soil. The larger size and fewer the number of pores the easier it is for water to move through the soil profile.
It is not just the soil type that affects structure and drainage but also the activities or environmental factors occurring to them. Root and earthworm activity can improve soil structure through creating large pores. Excessive cultivation, removal of crop residues and increased traffic contribute to soil structural decline, through compaction of soils, reducing pore size and breaking down of soil aggregates.

The chemical make-up of soils also determines structure. When high amounts of sodium are present (>6% exchangeable sodium percentage) clay particles separate and move freely about in wet soil. These soils are known as sodic soils. When sodic soils come in contact with water, the water turns milky as the clay disperses and when the soil dries out a crust is formed on the surface. Sodicity can be overcome by applying gypsum.

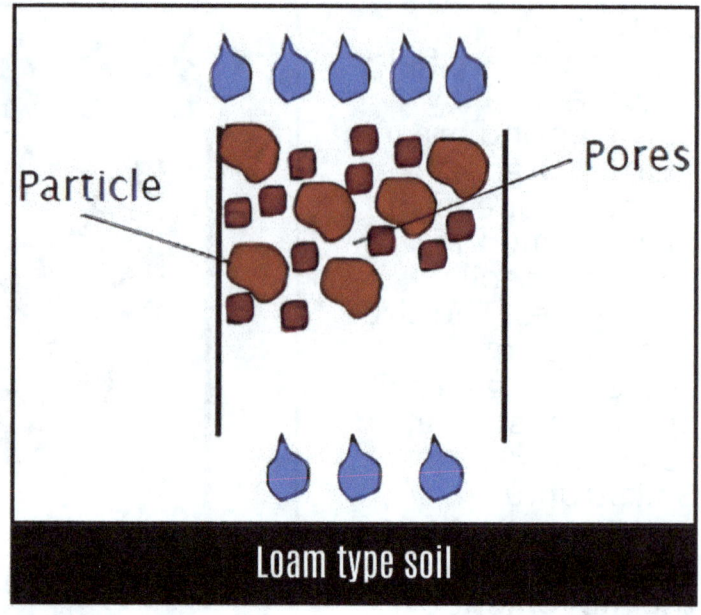

Slaking is the breakdown of aggregates on wetting, into smaller particles. Slaking generally occurs when intense rainfall hits dry soil, the aggregates collapse as a result of the pressure created by the clay swelling and the trapped air expanding and escaping. This process can block up pore spaces and when the soil dries a crust is formed, causing infiltration and seedling emergence problems.

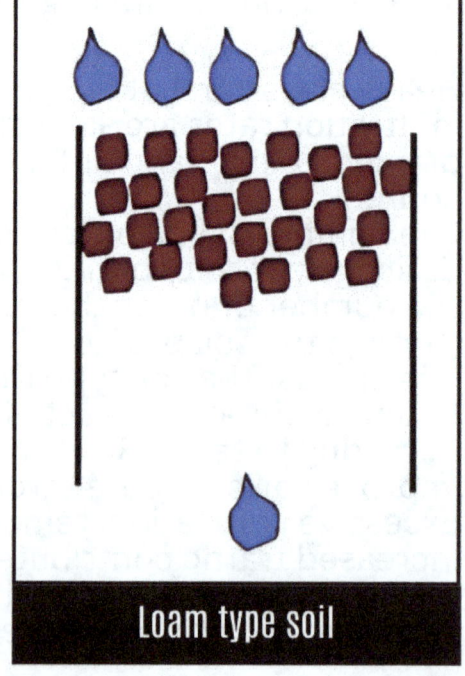

Soil colour

Soil colour can indicate the organic matter content of soil, the parent material soil is formed from, the degree of weathering the soil has undergone and the drainage characteristics of the soil.
The colour of the soil is the main indicator of how soils drain.

Table 1: Soil colour and indications

SOIL COLOUR		INDICATION
Dark brown		High organic matter content
Black		Humus
Red		Presence of ironPhosphorous may be less available to the plantFree draining

Yellow		Moist conditionsRestrictive drainageLess weathering
Grey, Blue/green hues		Poor drainageWaterlogging

Lighter coloured soils can generally indicate low fertility, for example, white sands. While darker soils (like black clays) are quite fertile. There is a large range in between.

Determine soil drainage

The drainage of a soil is an important characteristic to assess, as many plants prefer well-drained soils.
If a soil is poorly drained, sufficient oxygen cannot get to the plant roots, which can stunt or kill the plant. Soils that are very well drained can limit plant capture of water in drier environments or in dry years due to insufficient water holding capacity.
Other important indicators are:
- texture of the soil
- presence of buckshot and stones
- dispersibility and friability of the soil.

Inorganic component of soils

Inorganic material is the major component of most soils. It consists largely of mineral particles with specific physical and chemical properties which vary depending on the parent material and conditions under which the soil was formed.

It is the inorganic fraction of soils which determines soil physical properties such as texture. This has a large effect on structure, density and water retention.

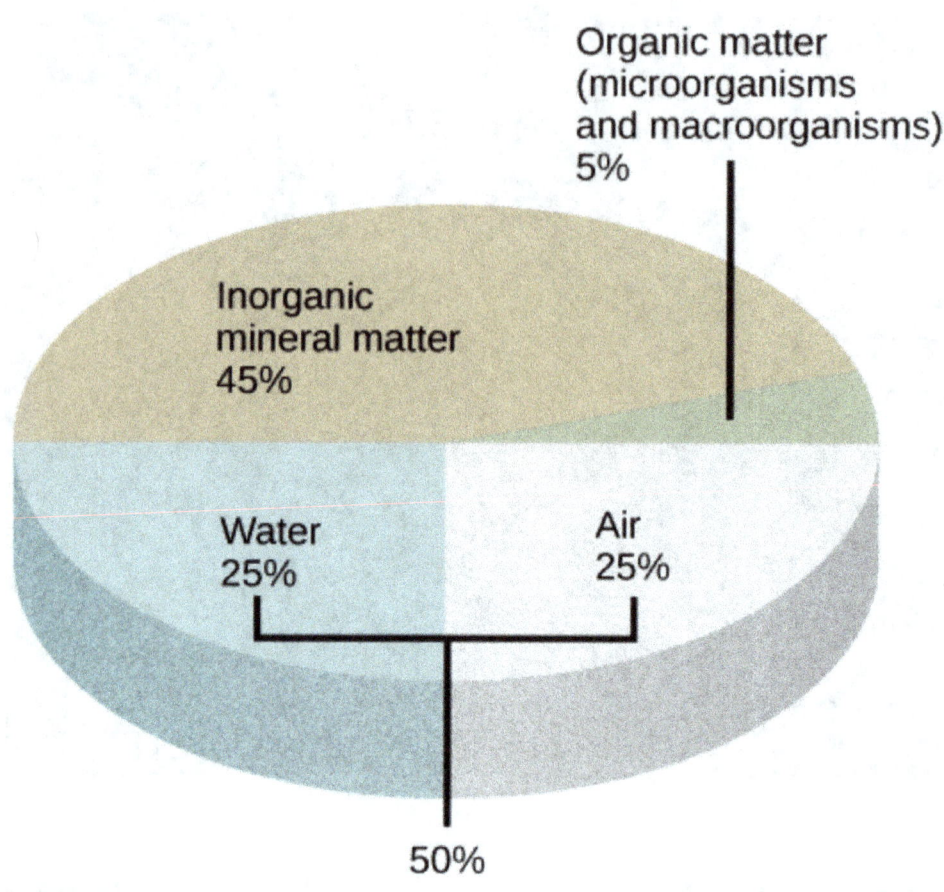

Soil texture

The texture of soil is a property which is determined largely by the relative proportions of inorganic particles of different sizes.
In Australia, the following sizes are used to describe the inorganic fraction of soils:
1. **Gravel** — particles greater than 2mm in diameter
2. **Coarse sand** — particles less than 2mm and greater than 0.2mm in diameter

3. **Fine sand** — particles between 0.2mm and 0.02mm in diameter
4. **Silt** — particles between 0.02mm and 0.002mm in diameter
5. **Clay** — particles less than 0.002mm in diameter.

SAND

Quartz is the predominant mineral in the sand fraction of most soils. Sand particles have:
- a relatively small surface area per unit weight
- low water retention
- little chemical activity compared with silt and clay.

SILT

Silt has a relatively limited surface area with little chemical activity. Soils high in silt may compact under heavy traffic. This affects the movement of air and water in the soil.

CLAY

Clays have very large surface areas compared with the other inorganic fractions. As a result, clays are chemically very active and able to hold nutrients on their surfaces. These nutrients can be released into soil water to be used by plants. Like nutrients, water also attaches to the surface of clay but this water can be hard for plants to use.

There are many different types of clays. Clays are distinguished from sand and silt by their ability to swell and retain a shape they have been formed into — as well as by their sticky nature.

Soil textural class

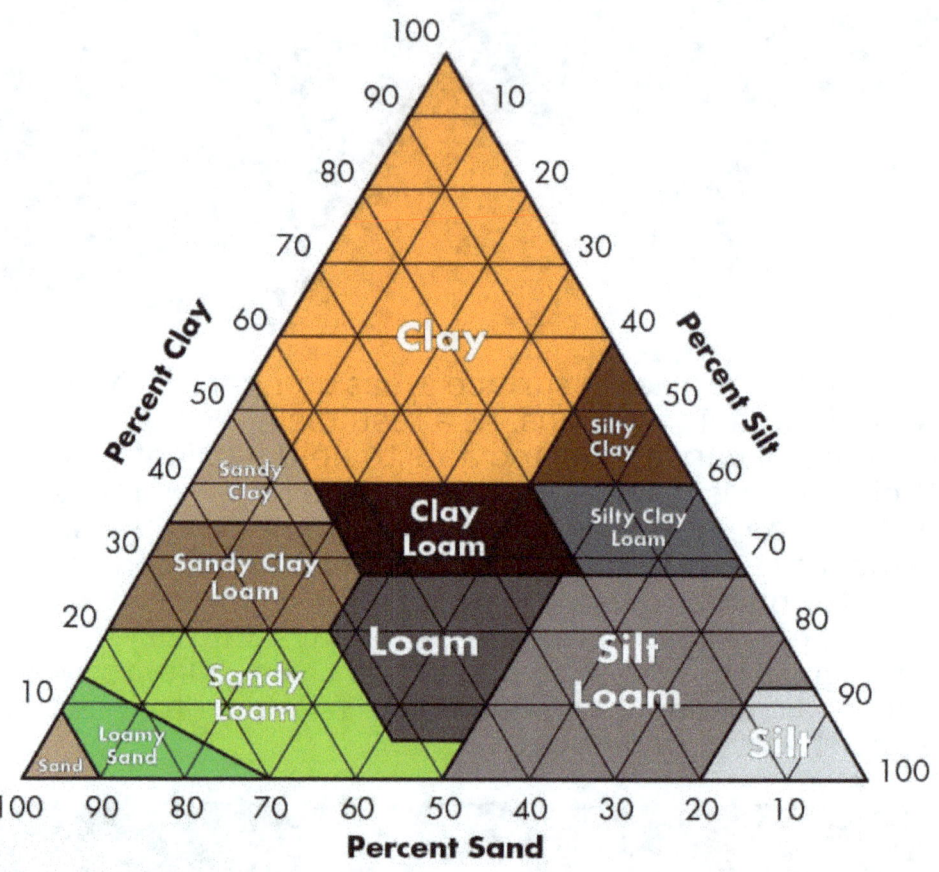

The relative proportion of sand, silt and clay particles determines the physical properties of soil, including the texture. The surface area of a given amount of soil increases significantly as the particle size decreases. Consequently, the soil textural class also gives an indication of soil chemical properties.
The exact proportions of sand, silt and clay in a soil can only be determined in a laboratory. However, a naming

system has been developed to approximately describe the relative proportions. This classification of soil can be undertaken in the field where particular properties indicate possible textural classes.

To estimate the texture in the field, crush a small sample of soil (10 to 20g) in one hand. After removing any gravel or root matter, work the soil in the fingers to break down any aggregates. With the sample moist but not sticky, the textural class can be estimated by the feel of the sample between the fingers.

TEXTURAL CLASS DESCRIPTIONS FOR SOIL

A simple way to determine a soil texture and its characteristics is by hand texturing. When texturing soil it is important to understand the behaviour feel, colour, sound and cohesiveness of the soil, which is achieved by making a bolus (wetting the soil and forming a ball). For example, a sandy loam will only just stick together (slightly coherent) and there will be noticeable sand grains which can be seen and felt and heard if you place the bolus close to your ear and squeeze it.

It is then important to form a ribbon from the bolus to determine the clay content of the soil. The longer the ribbon the higher the clay content. The length of the ribbon is measured against a ruler and along with the behaviour of the soil can be compared with the descriptions on the soil texture table. This table will help you to assess soil texture.

Table 2: Guide to common soil textures

Texture grade	Behaviour of moist bolus (ball formed in palm of hand)
Sand	Coherence, nil. Single sand grains adhere to fingers. If you press the bolus between your fingers, holding close to your ear, you will hear the sand grains rubbing against each other.
Loamy sand	Slight coherence. Discolours fingers with dark organic stain. Ribbon length 1.0cm.
Clayey sand	Slight coherence; sticky when wet. Many sand grains stick to fingers. Discolours fingers with clay stain. Ribbon length 1.0cm.
Sandy loam	Bolus just coherent but very sandy to touch. Ribbon length 1.3 to 2.5cm. Can hear sand grains (see Sand description).
Fine sandy loam	Bolus coherent. Sand can be felt and heard when manipulated. Ribbon length 1.3 to 2.5cm.

Texture grade	Behaviour of moist bolus (ball formed in palm of hand)
Light sandy clay loam	Bolus strongly coherent but sandy to touch. Ribbon length 2 to 2.5cm.
Loam	Bolus coherent and spongy. Smooth feel, may be greasy. Ribbon length 2.5cm.
Loam fine sandy	Bolus coherent and slightly spongy. Fine sand can be felt and heard when manipulated. Ribbon length 2.5cm.
Silt loam	Coherent bolus, very smooth to silky when manipulated. Ribbon length 2.5cm.
Sandy clay loam	Strongly coherent bolus sandy to touch. Medium sand grains visible. Ribbon length 2.5 to 3.8cm.
Clay loam	Coherent plastic bolus, smooth to manipulate. Ribbon length 4 to 5cm.
Silty clay loam	Coherent smooth bolus, plastic and silky to touch. Ribbon length 4 to 5cm.

Texture grade	Behaviour of moist bolus (ball formed in palm of hand)
Fine sandy clay loam	Coherent bolus, fine sand can be felt and heard. Ribbon length 4 to 5cm.
Sandy clay	Plastic bolus, fine medium sands can be seen, felt or heard in clay matrix. Ribbon length 5 to 7.5cm.
Silty clay	Plastic bolus, smooth and silky to manipulate. Ribbon length 5 to 7.5cm.
Light clay	Plastic bolus, smooth to touch; slight resistance to shearing between thumb and forefinger. Ribbon length 5 to 7.5cm.
Light medium clay	Plastic bolus, smooth to touch, slightly greater resistance to ribboning. Ribbon length 7.5cm.
Medium clay	Smooth plastic bolus, handles like plasticine. Has some resistance to ribboning. Ribbon length 7.5cm.
Heavy clay	Smooth plastic bolus, handles like stiff plasticine. Has firm resistance to ribboning. Ribbon length 7.5cm or more.

It should always be remembered that soil texture often varies with depth and that the properties of the topsoil are affected by the properties of the subsoil.

STRUCTURE

Structure is the arrangement of soil particles and pore spaces between. Soil with a structure beneficial to plant growth, has stable aggregates between 0.5 and 2mm in diameter. Such soils have good aeration and drainage.

CHEMICAL PROPERTIES

The inorganic minerals of soils consist primarily of silicon, iron and aluminium which do not contribute greatly to the nutritional needs of plants. Those in the clay fraction have the capacity to retain nutrients in forms which are potentially available for plants to use.

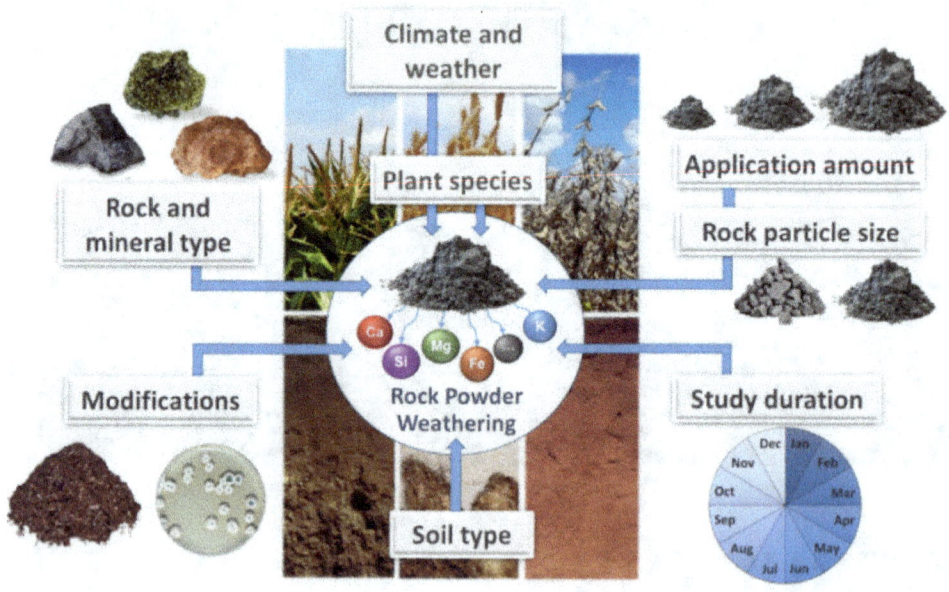

Organic component of soil

The organic matter of soil usually makes up less than 10% of the soil. It can be subdivided into living and the non-living fractions. The non-living fraction contributes to the soil's ability to retain water and some nutrients and to the formation of stable aggregates.

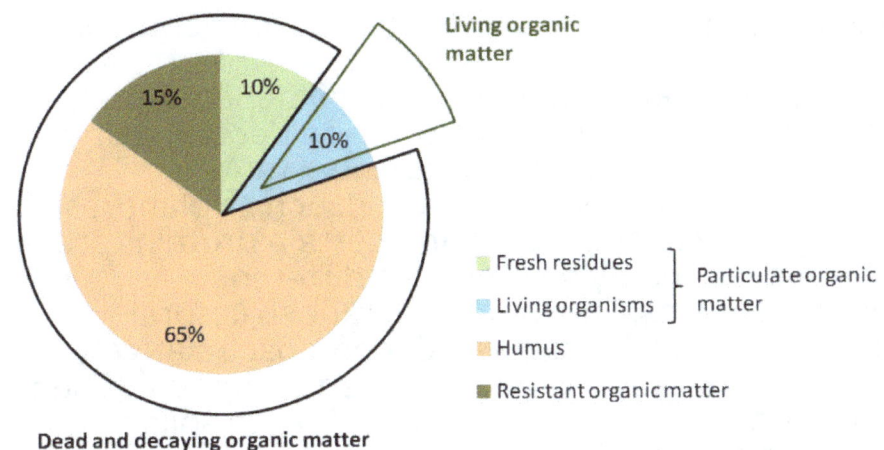

Dead and decaying organic matter

ORGANIC MATTER FRACTION OF SOILS

The organic matter fraction of soils comes from the decomposition of animal or plant products such as faeces and leaves. Soil organic matter contributes to stable soil aggregates by binding soil particles together. Plants living in soil continually add organic matter in the form of roots and debris. Decomposition of this organic matter by microbial activity releases nutrients for the growth of other plants.

The organic matter content of a soil depends on the rates of organic matter addition and decomposition. Soil microorganisms are responsible for the decomposition of organic matter such as plant residues. Initially, the sugars, starch and certain proteins are readily attacked by a number of different microorganisms. The more resistant structural components of the cell wall decompose relatively slowly. The less easily decomposed compounds, such as lignin and tannin, impart a dark colour to soils containing a significant organic matter content.

The decomposition rate of organic materials depends on how favourable the soil environment is for microbial activity.

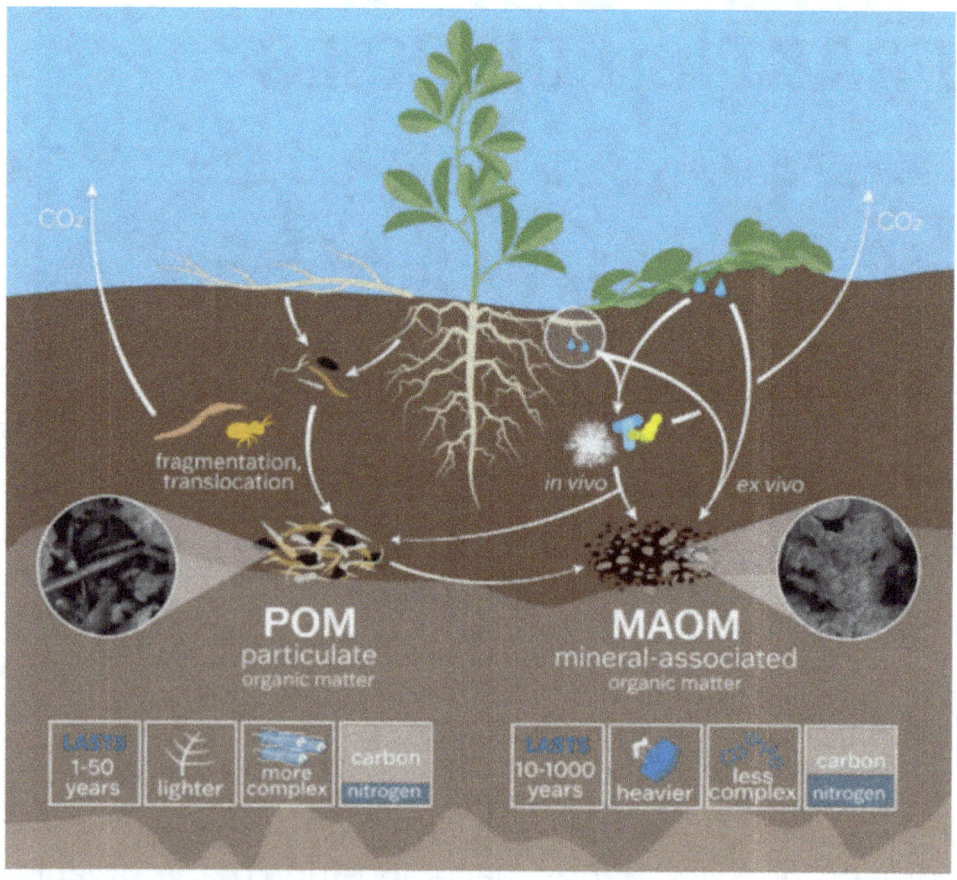

Higher decomposition rates occur where there are:
- warm, moist conditions
- good aeration
- a favourable ratio of nutrients
- a pH near neutral
- freedom from toxic compounds.

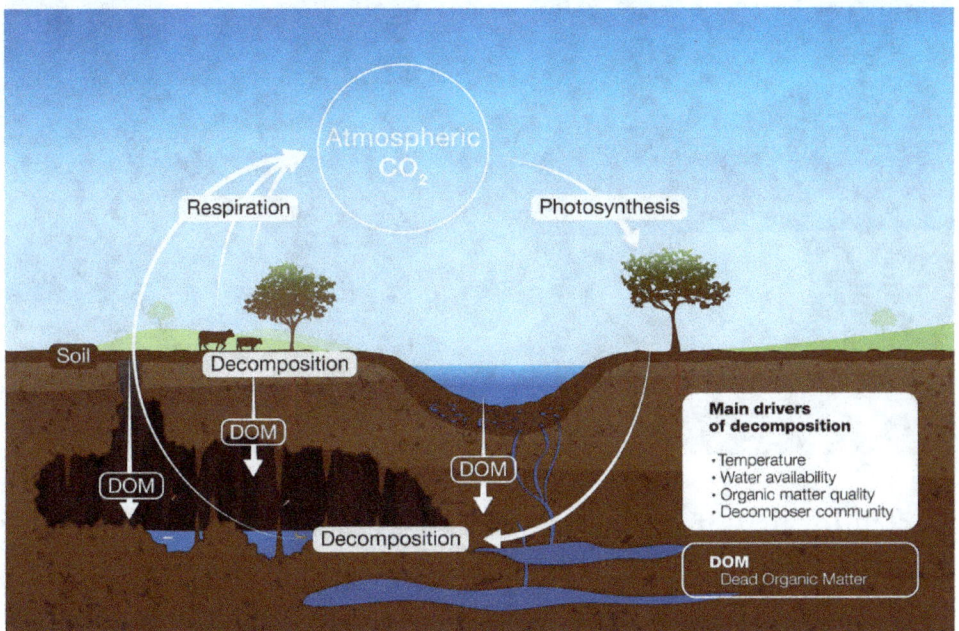

Soil organisms

The soil contains numerous organisms ranging from microscopic bacteria to large soil animals such as earthworms. The soil microorganisms include:
- bacteria
- fungi
- actinomycetes
- algae
- protozoa
- nematodes.

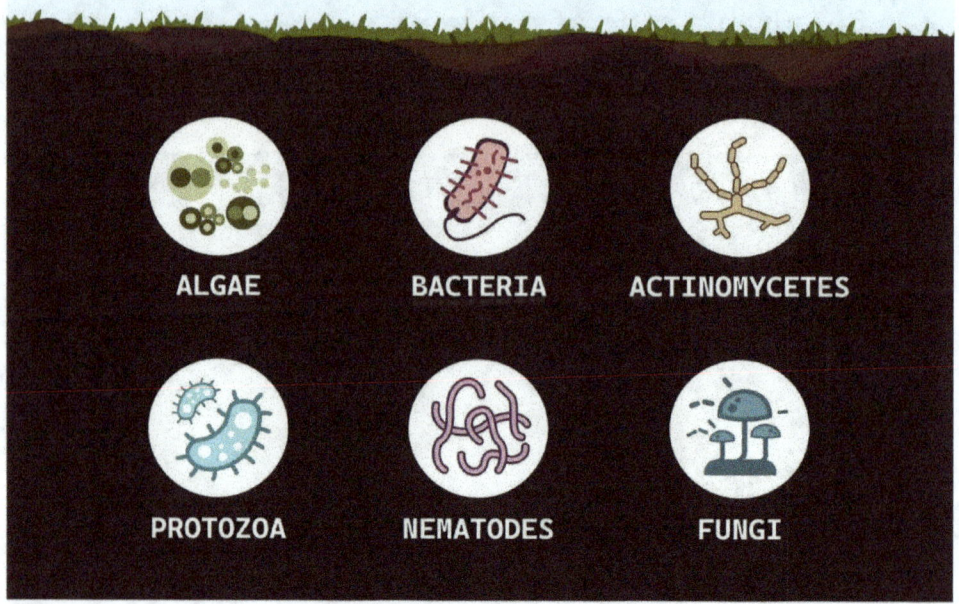

The diversity of soil organisms can both assist and hinder plant growth. Beneficial activities include:
- organic matter decomposition
- nitrogen fixation
- transformation of essential elements from one form to another
- improvement in soil structure through soil aggregation
- improved drainage and aeration.

Under some circumstances, soil organisms compete with plants for nutrients.

Bacteria are the smallest and most numerous microorganisms in the soil. They make an important contribution to organic matter decomposition, nitrogen fixation and the transformation of nitrogen and sulphur. The fungi and actinomycetes contribute beneficially to organic matter decomposition. The group of large soil

animals includes earthworms, which incorporate organic matter into the soil as well as improve aeration and drainage by means of their channels.
Some soil fungi, nematodes, and insects feed on roots and lateral shoots to the detriment of plants.

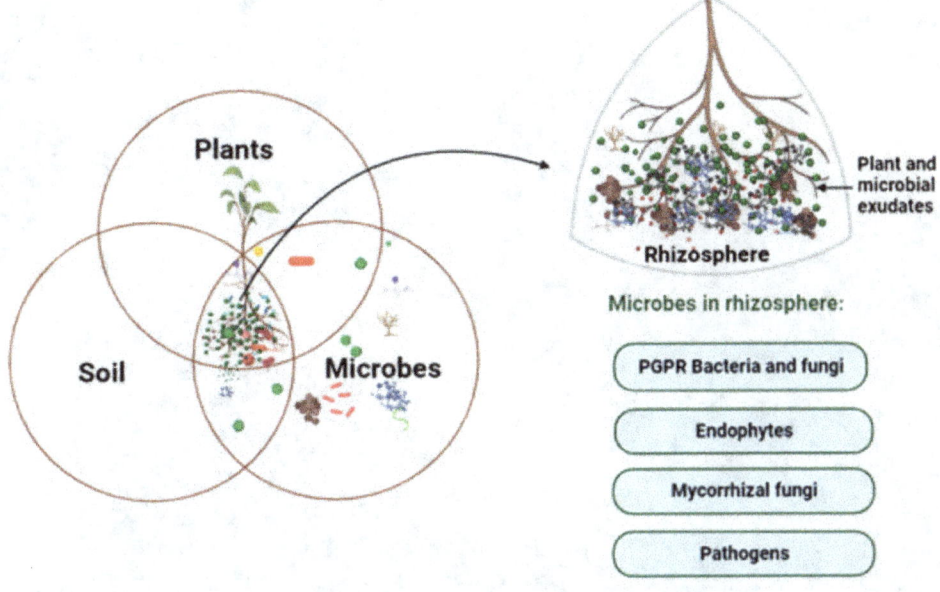

Soil acidity

Soil acidity is a potentially serious land degradation issue. When soil becomes too acidic it can:
- decrease the availability of essential nutrients
- increase the impact of toxic elements
- decrease plant production and water use
- affect essential soil biological functions like nitrogen fixation
- make soil more vulnerable to soil structure decline and erosion.

Without treatment, soil acidification can impact agricultural productivity and sustainable farming systems. Acidification can also extend into subsoil layers, posing serious problems for plant root development and remedial action.

What causes soil acidity?

Soil acidification is a natural process, but it can be increased by some agricultural practices. Acidification occurs in agricultural soils as a result of the:
- removal of plant and animal products
- leaching of excess nitrate
- addition of some nitrogen based fertilisers
- build-up in mostly plant-based organic matter.

Soil acidity occurs naturally in higher rainfall areas and can vary according to:
- the landscape geology
- clay mineralogy
- soil texture
- buffering capacity.

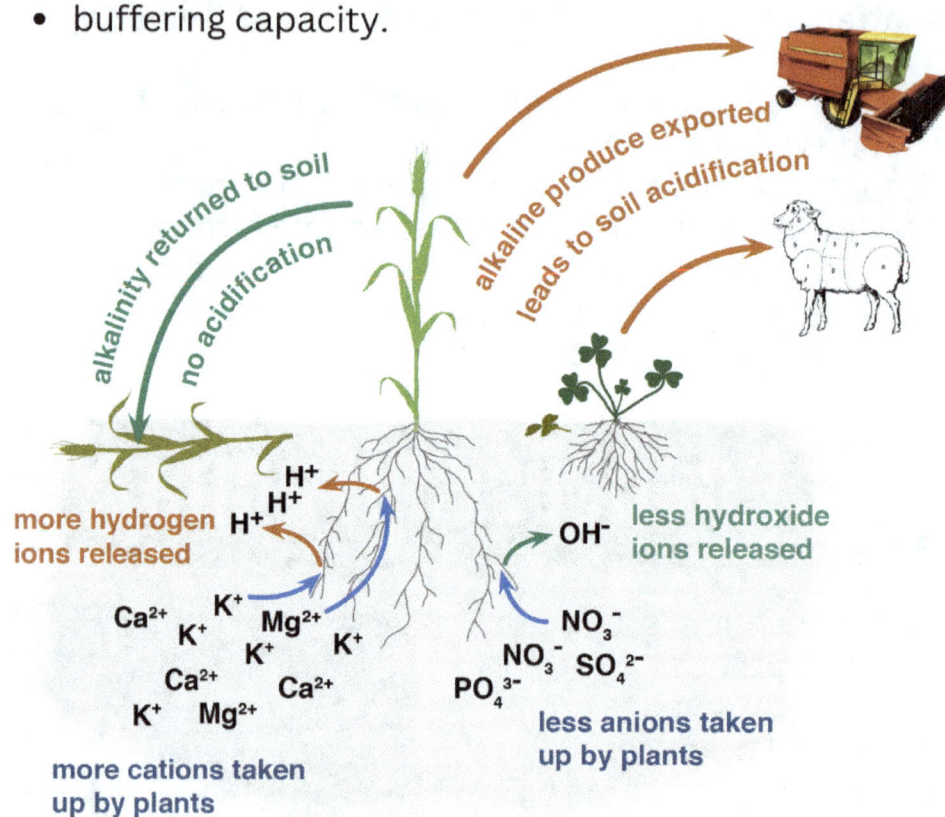

How acidity affects plant growth

Acidity itself is not responsible for restricting plant growth. Instead, biological processes favourable to plant growth can be negatively affected by acidity.

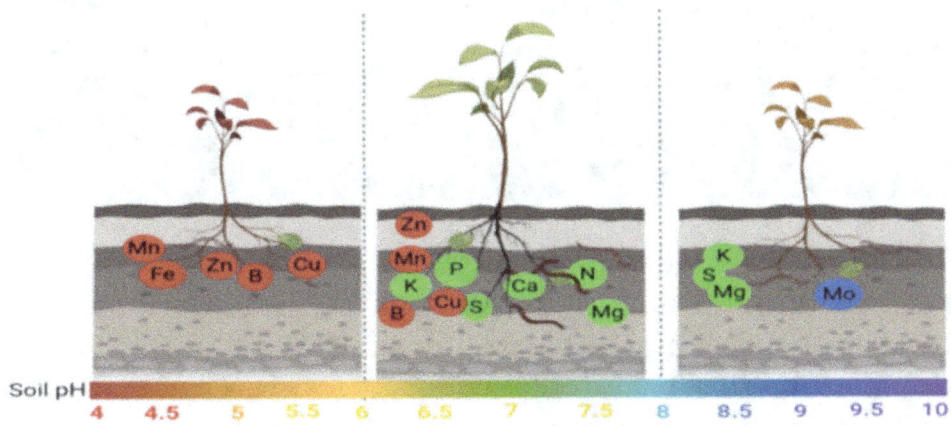

Bioavailibility of plant nutrient to soil pH

Acidity has the following effects on soil:
- It decreases the availability of plant nutrients, such as phosphorus and molybdenum, and increases the availability of some elements to toxic levels, particularly aluminium and manganese.
- Essential plant nutrients can also be leached below the rooting zone.
- Acidity can degrade the favorable environment for bacteria, earthworms and other soil organisms.
- Highly acidic soils can inhibit the survival of useful bacteria, such as the rhizobia bacteria that fix nitrogen for legumes.

The rhizobia

Soil pH as a measure of acidity

Soil pH is a measure of acidity or alkalinity. A pH of 7 is neutral, above 7 is alkaline and below 7 is acid. Because pH is measured on a logarithmic scale, a pH of 6 is 10 times more acid than a pH of 7.
Soil pH can be measured either in water (pHw) or in calcium chloride (pHCa) and the pH will vary depending on the method used. As a general rule, pH measured in calcium chloride is 0.7 of a pH unit lower than pH measured in water.

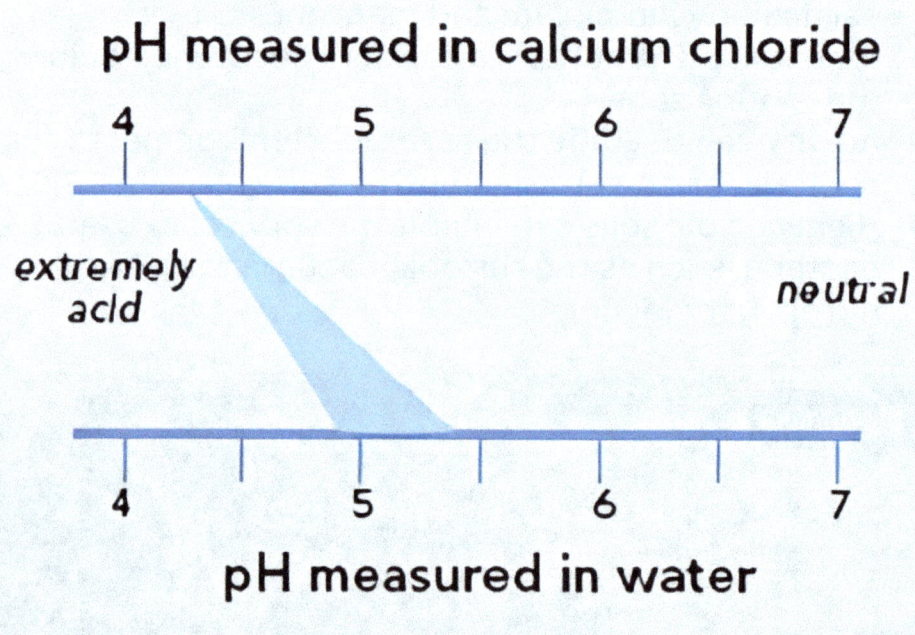

There are several differences between pHCa and pHw:
- Soil pHCa measurements in Australia vary from pHCa 3.6 to pHCa 8 for a range of different soil textures (sandy loams to heavy clays). Soil pHw values lie between pHw 4 and pHw 9.
- The pHw may be higher by 0.6 to 1.2 in low salinity soils and higher by 0.1 to 0.5 in high salinity soils. Research has shown a difference of 0.7 for a wide range of soils.
- Higher pHw values to around 10 may be associated with alkali mineral soils containing sodium carbonates and bicarbonates.
- Research has shown that seasonal variation of pHw can vary up to 0.6 of a pH unit in any one year. In comparison, soil pHCa measurements are less affected by seasons.

When a laboratory measures your soil's pH, make sure they specify which method (water or calcium chloride) was used.

SOIL PH LEVELS

A pHCa range between 5 and 6 is considered ideal for most plants. Acid soils have a major effect on plant productivity once the soil pHCa falls below 5:

- pH 6.5 — close to neutral — Optimum for many acid-sensitive plants. Some trace elements may become unavailable.
- pH 5.5 — slightly acid — Optimal balance of major nutrients and trace elements available for plant uptake.
- pH 5.0 — moderately acid — Below pH 4.8 aluminium (Al) can become toxic to plants, depending on soil type. Phosphorus combines with Al and may be less available to plants.
- pH 4.5 — strongly acid — Aluminium becomes soluble in toxic quantities. Manganese (Mn) becomes soluble and toxic to plants in some soils, depending on temperature and moisture conditions. Molybdenum (Mo) is less available. Soil bacterial activity is slowed down.
- pH 4.0 — extremely acid — Irreversible soil structural breakdown can occur.

Soil pH will influence both the availability of soil nutrients to plants and how the nutrients react with each other.

For example:

- At a low pH, many elements become less available to plants, while others such as iron, aluminum and manganese become toxic to plants. Aluminum, iron and phosphorus also combine to form insoluble compounds.
- At a high pH, calcium ties up phosphorus, making it unavailable to plants, and molybdenum becomes toxic in some soils. Boron may also be toxic in some soils.

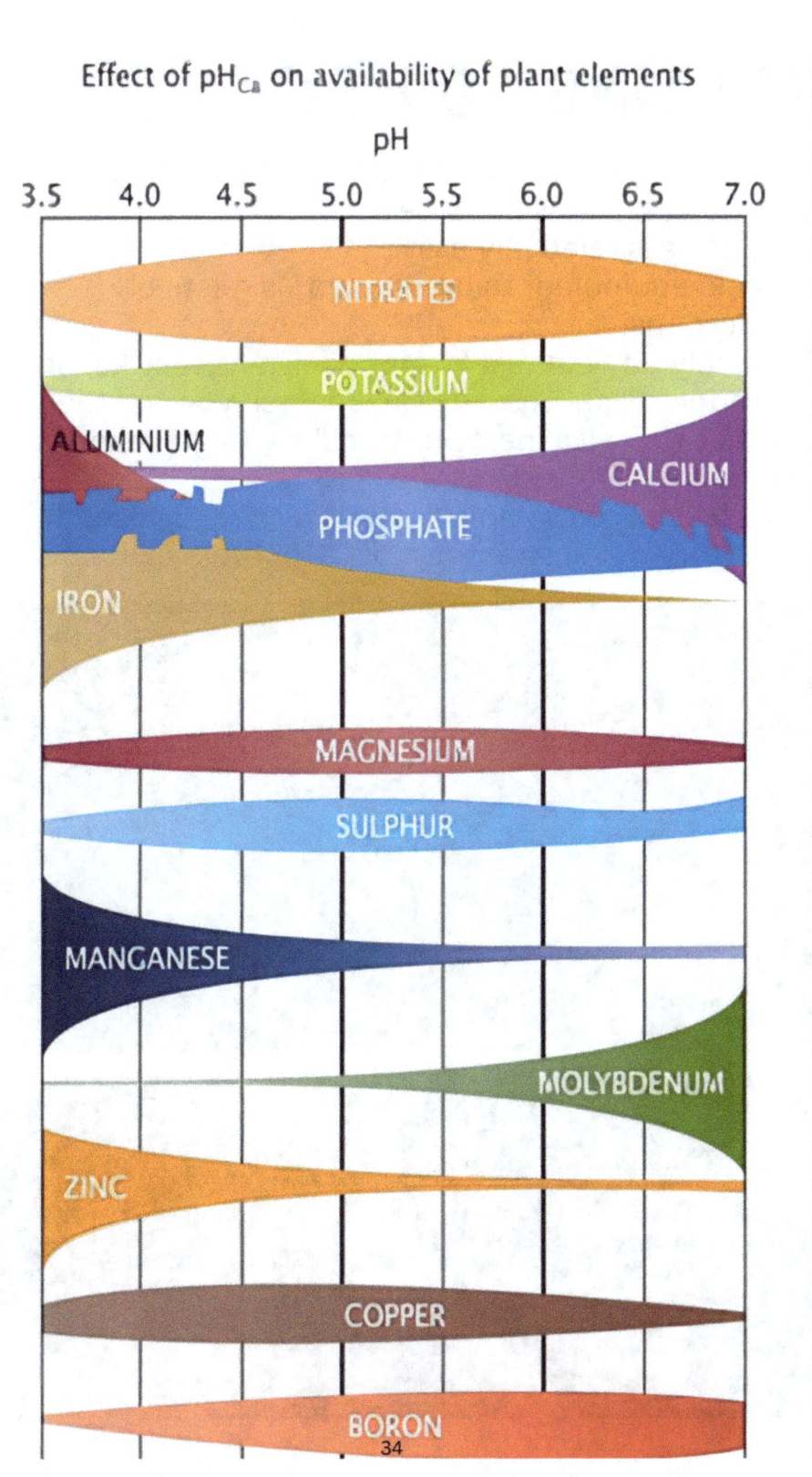

TESTING SOIL PH

Soil pH is one of the most routinely measured soil parameters. This is because:
- testing is relatively easy
- field equipment to measure pH is relatively inexpensive.

Don't rely on field test kits for decisions such as rates of lime application. Test kits will only tell you whether your soil is acid or alkaline. You're unlikely to get responses to lime if other nutrients are lacking.

Professional soil sample analysis by a recognised laboratory will ensure the most accurate results.

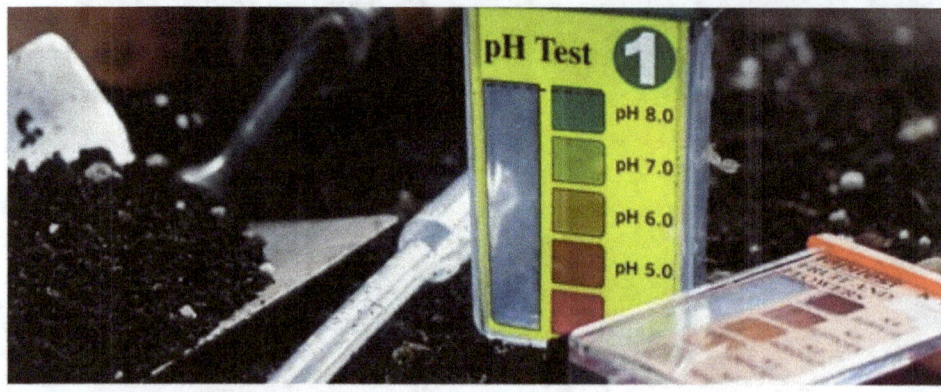

LIMING TO CORRECT SOIL PH

For most acid soils, the most practical management option is to add lime to maintain the current soil pH status or increase surface soil pH.

For a better chance at successfully growing acid-sensitive species, consider liming once the pH drops below pHCa 5.0.

If paddocks with an acidity problem are not limed, the soil pH will continue to fall and settle at pHCa 3.8 to 4.2.

LIME APPLICATION FOR PERMANENT PASTURE

For most acid soils, the most practical management option is to add lime to maintain the current soil pH status or increase surface soil pH.

For a better chance at successfully growing acid-sensitive species, consider liming once the pH drops

below pHCa 5.0.

If paddocks with an acidity problem are not limed, the soil pH will continue to fall and settle at pHCa 3.8 to 4.2.

Composting for Soil

How to make nutrient-rich, garden 'gold' that will help your garden thrive.

Composting benefits

- Excellent soil conditioner: By making compost, you are creating rich humus for your lawn and garden. This adds nutrients to your plants and helps retain soil moisture.
- Recycles kitchen and yard waste: Composting can divert as much as 30% of household waste away from the garbage can. That's important, because when organic matter hits the landfill, it lacks the air it needs to decompose quickly. Instead, it creates harmful methane gas as it breaks down, increasing the rate of climate change.
- Introduces beneficial organisms to the soil: Microscopic organisms in compost help aerate the soil, break down organic materials for plant use, and ward off plant disease.
- Good for the environment: Composting offers a natural alternative to chemical fertilizers when applied to lawns and garden beds.

What to compost

All compostable materials are either carbon or nitrogen-based, to varying degrees. The secret to a healthy compost pile is to maintain a working balance between these two elements.

Carbon
Carbon-rich matter (like branches, stems, dried leaves, peels, bits of wood, sawdust or shredded paper) gives compost its light, fluffy body.

Nitrogen
Nitrogen- or protein-rich matter (manures, food scraps, green lawn clippings, kitchen waste, and green leaves) provides raw materials for making enzymes.

A healthy compost pile should have much more carbon than nitrogen, but since most materials are not pure carbon or nitrogen, a simple rule of thumb is to use one-third green materials and two-thirds brown. The bulkiness of the brown materials allows oxygen to penetrate and nourish the organisms that reside there. Too much nitrogen makes for a dense, smelly, slowly decomposing anaerobic mass. Good hygiene means covering fresh nitrogen-rich material with carbon-rich material, which often exudes a fresh, wonderful smell. If in doubt, add more carbon!

The table below details how the items in your compost are likely to be classified.

Material	Carbon/Nitrogen	Information
Wood chips / pellets	Carbon	High carbon levels; use sparingly
Wood ash	Carbon	Only use ash from clean materials; sprinkle lightly
Tea leaves	Nitrogen	Loose or in bags
Table Scraps	Nitrogen	Add with dry carbon items
Straw or hay	Carbon	Straw is best; hay (with seeds) is less ideal
Shrub prunings	Carbon	Woody prunings are slow to break down
Shredded paper	Carbon	Avoid using glossy paper and colored inks
Seaweed and kelp	Nitrogen	Apply in thin layers; good source for trace minerals

Material	Carbon/Nitrogen	Information
Sawdust pellets	Carbon	High carbon levels; add in layers to avoid clumping
Pine needles	Carbon	Acidic; use in moderate amounts
Newspaper	Carbon	Avoid using glossy paper and colored inks
Leaves	Carbon	Leaves break down faster when shredded
Lawn & garden weeds	Nitrogen	Only use weeds which have not gone to seed
Green comfrey leaves	Nitrogen	Excellent compost 'activator'
Grass clippings	Nitrogen	Add in thin layers so they don't mat into clumps
Garden plants	--	Use disease-free plants only
Fruit and vegetable scraps	Nitrogen	Add with dry carbon items

Material	Carbon/Nitrogen	Information
Flowers, cuttings	Nitrogen	Chop up any long woody stems
Eggshells	Neutral	Best when crushed
Dryer lint	Carbon	Best if from natural fibers
Corn cobs, stalks	Carbon	Slow to decompose; best if chopped up
Coffee grounds	Carbon	Filters may also be included
Chicken manure	Nitrogen	Excellent compost 'activator'
Cardboard	Carbon	Shred material to avoid matting

WHAT NOT TO COMPOST

- Do not compost meat, bones, or fish scraps (they will attract pests) unless you are using a composter designed specifically for this purpose. The Green Cone Solar Waste Digester or the Jora compost tumbler are two examples of composters that will accommodate these materials.
- Avoid composting perennial weeds or diseased plants, since you might spread seeds or diseases.
- Don't include pet manures in compost that will be used on food crops.

- Banana peels, peach peels, and orange rinds may contain pesticide residues and should be kept out of the compost.
- Black walnut leaves should not be composted.
- Sawdust may be added to the compost, but should be mixed or scattered thinly to avoid clumping. Be sure sawdust is clean, with no machine oil or chain oil residues.

✓ YES	NO ✗
Veggie & fruit scraps	Meat and bones
Coffee & tea grounds	Dairy products
Dryer lint	Pet waste
Egg shells	Fats & oil
Leaves & plant material	Medication
Shredded paper	Weeds with seeds
Grass clippings	Diseased plants

How to compost step-by-step

1. Start your compost pile on bare earth. This allows worms and other beneficial organisms to aerate the compost and be transported to your garden beds.
2. Lay twigs or straw first, a few inches deep. This aids drainage and helps aerate the pile.
3. Add compost materials in layers, alternating moist and dry. Moist ingredients are food waste, tea bags, seaweed, etc. Dry materials are straw, leaves, sawdust pellets and wood ashes. If you have wood ashes, sprinkle in thin layers, or they will clump together and be slow to break down.
4. Add manure, green manure (clover, buckwheat, wheatgrass, grass clippings) or any nitrogen source. This activates the compost pile and speeds the process along.
5. Keep compost moist. Water occasionally, or let rain do the job.
6. Cover with anything you have – wood, plastic sheeting, carpet scraps. Covering helps retain moisture and heat, two essentials for compost. Covering also prevents the compost from being over-watered by rain. The compost should be moist, but not soaked and sodden.
7. Turn. Every few weeks give the pile a quick turn with a pitchfork or shovel. This aerates the pile. Oxygen is required for the process to work, and turning "adds" oxygen. You can skip this step if you have a ready supply of coarse material like straw. Once you've

established your compost pile, add new materials by mixing them in, rather than by adding them in layers. Mixing, or turning, the compost pile is key to aerating the composting materials and speeding the process to completion.

SIMPLEST COMPOSTING METHODS

"No-turn" composting

The biggest chore with composting is turning the pile from time to time. However, with 'no-turn composting', your compost can be aerated without turning.
The secret is to thoroughly mix in enough coarse

material when building the pile. The compost will develop as fast as if it were turned regularly, and studies show that the nitrogen level may be even higher than with turned compost.

With 'no-turn' composting, add new materials to the top of the pile and harvest fresh compost from the bottom of the bin.

Enclosed compost bins

For small-scale outdoor composting, enclosed bins are the most practical. Enclosed bins include:

- **DIY compost bin**: The least expensive method is to build one yourself from a heavy-duty garbage can. Drill 1.5-cm aeration holes in rows at roughly 15-cm intervals around the can. Fill with a mixture of high-carbon and high-nitrogen materials (see our table above). Stir the contents occasionally to avoid anaerobic pockets and to speed up the composting process. If the lid is secure, lay the can on its side and roll. A length of 2 x 2 cedar can be bolted to the inside, running top to bottom, to help flip the material. Without this, the contents tend to stay in place while the bin is rolled. Another option is to build your own bin from scrap lumber or spare wooden pallets.

- **Standard compost bins:** Another option is a compost bin, sometimes called a 'compost digester'. Compost bins are enclosed on the sides and top, and open on the bottom so they sit directly on the ground. These are common composting units for homes in residential areas where bins tend to be smaller, yet enclosed enough to discourage pests.

- **Tumblers:** The most efficient enclosed bin method is the compost tumbler. It's possible to maintain relatively high temperatures in drum/tumbler systems, because the container acts as insulation and the turning keeps the microbes aerated and active. Some designs help bring air into the compost and prevent clumping of the composting materials.

TIPS FOR SUCCESSFUL COMPOSTING

Activate your compost
'Activators' can be added to your compost to help kick-start decomposition and speed up composting. Common compost activators include: comfrey leaves, grass clippings, young weeds, and well-rotted chicken manure. You can also buy inoculant at your local garden center, though a shovel full of finished compost from another pile works just as well.

Minimize flying insects
Small fruit flies are naturally attracted to the compost pile. Discourage them by covering any exposed fruit or vegetable matter. Keep a small pile of grass clippings next to your compost bin, and when you add new kitchen waste to the pile, cover it with one or two inches of clippings. Adding lime or calcium will also discourage flies.

Minimize odors

First, remember to not put bones or meat scraps into the compost unless your composter can handle these ingredients.Second, cover new additions to the compost pile with dry grass clippings or similar mulch. Adding lime or calcium will also neutralize odors. If the compost smells like ammonia, add carbon-rich elements such as straw, peat moss or dried leaves.

Is your compost pile soggy?
This is a common problem, especially in winter, when carbon-based materials are in short supply. To solve this problem, you'll need to restore your compost to a healthy nitrogen-carbon balance.

Matted leaves and grass clippings clumping together?
This is a common problem with materials thrown into the composter. The wet materials stick together and slow the aeration process. There are two simple solutions: either set these materials to the side of the composter and add them gradually with other ingredients, or break them apart with a pitchfork. Grass clippings and leaves should be mixed with rest of the composting materials for best results.

Take advantage of autumn's bounty
The biggest challenge for small-scale backyard composting is finding enough carbon-rich materials to balance the regular input of nitrogen-rich materials

from kitchen scraps. Enter autumn leaves! These carbon-rich wonders are perfect for adding to your compost throughout the year. Just rake them up and save in bags placed near the compost for year-round contributions.

COMPOSTING WEED SEEDS?

A liability in composting is the unexpected introduction of new weed seeds to your garden. This is caused by slow or incomplete composting that didn't generate enough heat to kill weed seeds. Weed seeds in compost are a nuisance because once the compost is transferred to your garden beds, the compost acts to fertilize the weeds and make them even more persistent! With home compost bins or piles, the way to eliminate weed seeds is twofold:

- **Make sure your compost is hot enough.**

Specifically, the temperature should be 130 – 150 degrees F. It takes about 30 days at 140 degrees to kill weed seeds.

- **Mix your pile.**

While your compost may be hot in the center of the mass, the outside of the pile is cooler, giving seeds a chance to survive. Mixing brings cooler material to the warmer area and also increases aeration, which helps attain the higher heat levels. Compost tumblers are very useful for this.

A Step-by-Step Guide to Growing Microgreens and Herbs Indoors

It's well known that going vegan and increasing your intake of fresh vegetables and fruits can benefit both your physical and mental health and do a world of good for animals. One of the best parts about going vegan is learning new ways to use plants in your cooking, and there's nothing like having fresh greens and herbs on hand for your culinary adventures. Plus, eating plant-based foods means you won't be supporting highly abusive animal agriculture.

Planting microgreens and herbs is a great introduction to gardening—and it's easy enough that anyone can do it! In addition to being extremely nutritious and adding great flavor and texture to a dish, microgreens and herbs add a beautiful pop of color to your meal that will be sure to entice dinner guests. The best part is that you don't need lots of space, time, or money to grow them.

1. START BY CHOOSING YOUR SEEDS.

Microgreens are very young vegetables and herbs. Rather than allowing these plants to grow to maturity, you harvest them early—they're often ready to harvest only a few weeks after they've been planted. Some of the most popular varieties include radish greens, kale, arugula, peas, and basil.

When choosing which herbs you'd like to grow, consider what kinds of food you enjoy cooking and eating. If you often cook Italian dishes such as pizzas and pastas, you might enjoy having fresh basil and oregano on hand. Cilantro is wonderful in Mexican dishes, including tacos, burritos, and nachos. And lemon balm and peppermint are perfect for making your own medicinal teas. You can even grow catnip indoors for your cat to enjoy!

Common herbs you can grow indoors include the following:
- Basil
- Catnip
- Chives
- Cilantro
- Lemon balm
- Mint
- Oregano
- Sage
- Thyme

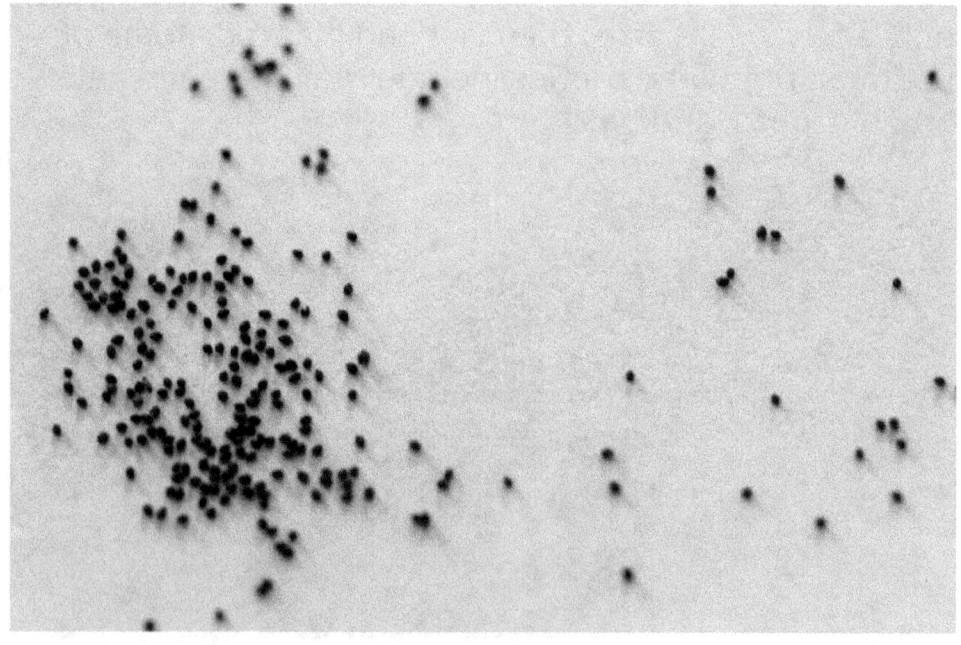

2. GATHER YOUR GROWING CONTAINERS.

For both herbs and microgreens, you'll need well-draining containers. Shallow trays are great for microgreens, as they have short roots. You can purchase specific trays just for microgreens or be even more eco-friendly by repurposing plastic food containers that you already have at home.

Herbs need deeper containers because their roots are longer. Options for these pots or containers are endless and can be made of plastic, metal, terracotta, or ceramic—just be sure that whatever container you choose has adequate drainage. Eight- to 10-inch pots are a good standard size for most herbs. If you get a larger pot, you can even plant multiple herbs in one container, as long as they have similar light, water, and nutrient requirements.

3. CHOOSE YOUR SOIL MIXTURE.

A good soil option for microgreens and herbs will contain a mixture of peat moss or coconut coir and perlite, all of which can be found at your local gardening store. You can also find indoor potting mix or seed starting mix that will work well for microgreens and herbs. Check the label for whatever soil mix you choose to make sure it doesn't include any animal-derived ingredients, manure, blood and bone meal from factory-farmed animals, or fish emulsion and fish meal from the rapacious commercial fishing industry.

4. PLANT YOUR HERBS AND MICROGREENS.

Each plant has a slightly different growing timeline and may require extra steps, such as soaking seeds before planting. Be sure to read the directions, usually printed on the back of your seed packet, before planting.

Microgreens

1. Prepare a draining container by filling it with your soil mix.
2. Sprinkle seeds over the potting mix and press them into the soil with your hands.
3. Cover the seeds to retain moisture, and then put them in a dark area, such as a cupboard.
4. Keep seeds in a dark space for the required amount of time. (The timeframe is typically listed on the seed packet.) Mist the soil regularly to keep it moist.
5. After the required time has passed, you can expose your microgreens—which should be little baby plants by now—to indirect sunlight.
6. After about a week, you can use scissors to harvest your microgreens by snipping just above the soil line. Most varieties regrow and can be cut several times, and the tray soil can be composted after use.

Herbs

You can either plant herbs from seeds or buy some that have already started growing. When planting herb seeds, be sure to read the directions on the seed packets. Each herb can require slightly different planting and care. In general, you will want to follow these steps:

1. Prepare your containers by filling them with soil.
2. Follow planting directions on the seed packet to determine seed planting depth.
3. After planting, find a sunny place to keep your herbs. You can also use grow lights if preferred, which is a good choice during the winter or if you have little access to natural light sources.
4. Keep the soil moist, and be patient as your herb grows.
5. To increase the lifetime of your plant, harvest only a small percentage of the leaves at once.

5. FOSTER THE RIGHT ENVIRONMENT FOR YOUR PLANTS TO THRIVE.

Most herbs thrive in temperatures between 65 and 70 degrees. They like sunlight! South-facing windows are great for herbs such as thyme, basil, and rosemary that like more sun. Others such as parsley and chives can do with a little less light and might prefer a window that faces east or west. If you want to grow them with a grow light, that's fine, too.

6. ENJOY USING YOUR HOMEGROWN HERBS AND MICROGREENS

After the excitement of planting and watching your herbs and microgreens grow, you get to enjoy them. Use them in cooking, for medicinal teas, in beeswax-free salves, and so on. Put basil on pastas, cilantro in a chimichurri sauce, and microgreens on top of a veggie protein bowl. Get creative and try some new vegan recipes with your fresh and homegrown herbs.

What Is the Best Soil to Use for Planting Flowers?

Soil falls into three main types - sand, clay and silt. Generally speaking, the best potting soil for growing flowers is an even mix of the three aforementioned soil types and is called sandy loam. This mix will ensure optimum growth conditions for most flowers. Yes, most flowers, but not all flowers. Depending on the plants you'll be looking after, you might need to go for a particular type of soil. So, let's take a look at the different types of soil and which plants grow best in which.

Types of soils and when to use them

Soil texture normally depends on the amount of silt, sand and clay it contains. That being said, the quality of nutrients and drainage properties of the soil will depend a lot on its texture.

Loam Soil

Loam soil consists of a well-balanced mix of sand, silt, clay and humus. The reasons this soil is so often used for growing plants are the following:

Advantages of loam soil

- It contains high calcium levels - Calcium is good for stimulating healthy plant growth. It helps maintain the balance of soil chemicals and ensures that water reaches the flowers' roots by improving the soil's water-retaining ability. Calcium also helps loosen the soil and this way, facilitating oxygen to reach the plant's roots. Calcium is great for reducing the amount of salt found in the soil. Which is great, considering that too much salt can damage a plant's root system and limit its ability to absorb all of the much needed nutrients.
- It has higher pH levels - Most plants prefer pH levels between 6.0 and 7.0. The plant's ability to grow is greatly influenced by the soil's pH level. Since loam soil's acidity is within the favorable range, it allows for good plant nutrients and other soil organisms, such as earthworms, to thrive in it.
- Has a gritty texture - Loam soil is dry and soft but has a nice gritty texture, which causes it to crumble easily. This type of texture helps increase its amazing draining properties, while also retaining water and plant nutrients. So, plants have consistent moisture and food. And since loam soil is crumbly, this also allows for air to easily flow through it all the way down to the roots.

Disadvantages of loam soil
- There are instances in which loamy soils contain stones that may affect the harvest of certain crops.

PLANTS THAT GROW WELL IN LOAMY SOIL

Dog's Tooth Violet
All species of dog's tooth violet prefer well-drained, loamy soil.

Rubus
This hardy plant prefers full to partial shade and grows best in moist loamy soil.

Wisteria
Wisterias love well-drained, fertile and rich in nutrients and organic matter loamy soil.

Silt Soil

Silt-based soil is composed of intermediate sized particles and can be a bit tricky to work with. Since there's a risk of it compacting when wet, you will need to increase its organic matter by mixing it in with compost and other soil microbe-rich products. Here are some of the advantages and disadvantages of using silt soil in your garden:

Advantages of silty soil
- Silt is very fertile and holds on to nutrients very well
- Has good water holding capacity

Disadvantages of silty soil
- Water filtration of silty soils is usually poor
- Can become hard and compact
- Silt soils often form a crust

PLANTS THAT GROW WELL IN SANDY SOIL

Yellow Iris
An adaptive flower, which is often grown around a garden pond or stream.

Swamp Milkweed
A flower, which thrives in soggy soil.

Japanese Iris
A flower that loves water and is best planted around a water feature or wet area inside your garden.

Sandy Soil

Sand is the most prevalent part of this type of soil and its consistency is light and gritty to the touch. Sandy soil doesn't have many benefits and can be used to grow only a few types of plants. However there are ways to make it more manageable. For example you can amend sandy soil with fewer fertilizers and small amounts of water but on a more regular basis. Its organic matter can be improved by adding compost to the mix.

Advantages of sandy soil
- Warms up quickly during spring

Disadvantages of sandy soil
- Dries out very quickly during summer
- Any nutrients you do put in it are oftentimes washed away during rainfall

PLANTS THAT GROW WELL IN SANDY SOIL

Butterfly Weed
This butterfly magnet of a plant loves full sun exposure and prefers well-draining, dry sand-based soil.

Adam's Needle
This plant loves dry sandy soil and hates damp soils, which cause its roots to rot. Also, Adam's Needle can tolerate salt spray, which is said to decrease blooming and enhance the plant's green colour.

Blanket Flower
This drought-tolerant plant thrives best in sandy, almost pH-neutral soil.

Wormwood
A perennial herb, which is drought-resistant and grows particularly well in less fertile, dry sandy soil.

Clay Soil

Clay soil contains copious amounts of clay and due to that, it drains very poorly. However, with the right management techniques you can improve its overall quality. Simply amend your clay soil with compost and products rich in soil microbes in order to improve its organic matter. Also, try not to work on the soil while it's still wet.

Advantages of clay soil
- Clay soils are good at holding onto nutrients
- This type of soil is amazing for growing plants which require copious amounts of water

Disadvantages of clay soil
- Drains very slowly
- Warms up slowly during spring
- Compacts easily
- Is often too alkaline

PLANTS THAT GROW WELL IN CLAY SOIL

Black-eyed Susan
This plant can survive in both loamy and clayey soils, but does need proper drainage.

Bee Balm
There are varieties of Bee Balm that prefer sandy soils and others that grow better in loamy or clayey types of soil.

Goldenrod
A very adaptable flower that can survive in pretty much any soil, even clay.

How to grow indoor plants with soil from your garden?

When it comes to growing houseplants, avoid scooping soil directly from your back garden. Since garden soil contains many different types of bacteria, this can be harmful for your houseplants.
However, if you're really keen on growing your houseplants in soil from your garden, make sure to sterilise it, first. Pasteurising outdoor soil will help you eliminate any disease, weeds or pest insects from it. The process itself is quite easy, simply spread your soil on a baking sheet, place it in the oven and bake for around

30 minutes at 180 °C. Just bear in mind that this method, although effective, will leave a foul odour in your kitchen once done.

Once the soil has been sterilised, you'll have to amend it with the necessary amount of sand and peat moss. Adding these to the mix will help improve your soil's drainage, air flow and moisture-retaining properties.

How to mix a batch of amazing soil for houseplants at home

Outdoor soil aside, another way to get great soil for your houseplants is to make your own mix. For this, you will need to buy a few different types of soil and soil

amendments. In order to get the best quality soil, buy and mix the following:
- Half a cubic yard of yard perlite
- Half a cubic yard of peat moss
- Ten pounds of bone meal
- Five pounds of limestone
- Five pound sof blood meal

Once you mix all five of these, store the finished product in a proper airtight container. Open it only when you need to use the soil.

THANKS FOR READING

www.ingramcontent.com/pod-product-compliance
Lightning Source LLC
Chambersburg PA
CBHW071951210526
45479CB00003B/892